Student Edition

Eureka Math

Grade 4
Modules 1 & 2

Special thanks go to the Gordon A. Cain Center and to the Department of Mathematics at Louisiana State University for their support in the development of *Eureka Math*.

For a free *Eureka Math* Teacher Resource Pack, Parent Tip Sheets, and more please visit www.Eureka.tools

Published by the non-profit Great Minds

Printed in the U.S.A.
This book may be purchased from the publisher at eureka-math.org
10 9 8 7 6 5 4 3 2

ISBN 978-1-63255-303-4

Name ______________________________ Date ______________

1. Label the place value charts. Fill in the blanks to make the following equations true. Draw disks in the place value chart to show how you got your answer, using arrows to show any bundling.

a. 10 × 3 ones = ________ ones = __________

b. 10 × 2 tens =_________ tens = _________

c. 4 hundreds × 10 = _________ hundreds = _________

2. Complete the following statements using your knowledge of place value:

   a. 10 times as many as 1 ten is ________tens.

   b. 10 times as many as _________ tens is 30 tens or ________ hundreds.

   c. ________________________________ as 9 hundreds is 9 thousands.

   d. _________ thousands is the same as 20 hundreds.

   Use pictures, numbers, or words to explain how you got your answer for Part (d).

3. Matthew has 30 stamps in his collection. Matthew's father has 10 times as many stamps as Matthew. How many stamps does Matthew's father have? Use numbers or words to explain how you got your answer.

4. Jane saved $800. Her sister has 10 times as much money. How much money does Jane's sister have? Use numbers or words to explain how you got your answer.

5. Fill in the blanks to make the statements true.

   a. 2 times as much as 4 is _______.

   b. 10 times as much as 4 is _______.

   c. 500 is 10 times as much as _______.

   d. 6,000 is ________________________________ as 600.

6. Sarah is 9 years old. Sarah's grandfather is 90 years old. Sarah's grandfather is how many times as old as Sarah?

Sarah's grandfather is _______ times as old as Sarah.

Name ______________________________ Date ______________

1. Label the place value charts. Fill in the blanks to make the following equations true. Draw disks in the place value chart to show how you got your answer, using arrows to show any regrouping.

   a. 10 × 4 ones = ________ ones = __________

   b. 10 × 2 tens = _________ tens = _________

   c. 5 hundreds × 10 = _________ hundreds = _________

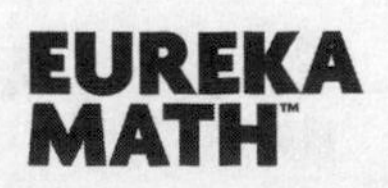

2. Complete the following statements using your knowledge of place value:

   a. 10 times as many as 1 hundred is ______ hundreds or ________ thousand.

   b. 10 times as many as _________ hundreds is 60 hundreds or ________ thousands.

   c. ______________________________ as 8 hundreds is 8 thousands.

   d. _________ hundreds is the same as 4 thousands.

   Use pictures, numbers, or words to explain how you got your answer for Part (d).

3. Katrina has 60 GB of storage on her tablet. Katrina's father has 10 times as much storage on his computer. How much storage does Katrina's father have? Use numbers or words to explain how you got your answer.

4. Katrina saved $200 to purchase her tablet. Her father spent 10 times as much money to buy his new computer. How much did her father's computer cost? Use numbers or words to explain how you got your answer.

5. Fill in the blanks to make the statements true.

   a. 4 times as much as 3 is _______.

   b. 10 times as much as 9 is _______.

   c. 700 is 10 times as much as _______.

   d. 8,000 is ________________________________ as 800.

6. Tomas's grandfather is 100 years old. Tomas's grandfather is 10 times as old as Tomas. How old is Tomas?

unlabeled thousands place value chart

This page intentionally left blank

Name ______________________________ Date ______________

1. As you did during the lesson, label and represent the product or quotient by drawing disks on the place value chart.

   a. 10 × 2 thousands = ________ thousands = ____________________

   b. 10 × 3 ten thousands = ________ ten thousands = ____________________

   c. 4 thousands ÷ 10 = ________ hundreds ÷ 10 = ____________________

2. Solve for each expression by writing the solution in unit form and in standard form.

Expression	Unit form	Standard Form
10 × 6 tens		
7 hundreds × 10		
3 thousands ÷ 10		
6 ten thousands ÷ 10		
10 × 4 thousands		

3. Solve for each expression by writing the solution in unit form and in standard form.

Expression	Unit form	Standard Form
(4 tens 3 ones) × 10		
(2 hundreds 3 tens) × 10		
(7 thousands 8 hundreds) × 10		
(6 thousands 4 tens) ÷ 10		
(4 ten thousands 3 tens) ÷ 10		

4. Explain how you solved 10 × 4 thousands. Use a place value chart to support your explanation.

5. Explain how you solved (4 ten thousands 3 tens) ÷ 10. Use a place value chart to support your explanation.

6. Jacob saved 2 thousand dollar bills, 4 hundred dollar bills, and 6 ten dollar bills to buy a car. The car costs 10 times as much as he has saved. How much does the car cost?

7. Last year the apple orchard experienced a drought and did not produce many apples. But this year, the apple orchard produced 45 thousand Granny Smith apples and 9 hundred Red Delicious apples, which is 10 times as many apples as last year. How many apples did the orchard produce last year?

8. Planet Ruba has a population of 1 million aliens. Planet Zamba has 1 hundred thousand aliens.
   a. How many more aliens does Planet Ruba have than Planet Zamba?

   b. Write a sentence to compare the populations for each planet using the words *10 times as many*.

G4-M1-SE-B1-1.3.1-1.2016

Name ______________________________ Date ______________

1. As you did during the lesson, label and represent the product or quotient by drawing disks on the place value chart.

   a. 10 × 4 thousands = ________ thousands = ____________________

   b. 4 thousands ÷ 10 = ________ hundreds ÷ 10 = ____________________

2. Solve for each expression by writing the solution in unit form and in standard form.

Expression	Unit Form	Standard Form
10 × 3 tens		
5 hundreds × 10		
9 ten thousands ÷ 10		
10 × 7 thousands		

3. Solve for each expression by writing the solution in unit form and in standard form.

Expression	Unit Form	Standard Form
(2 tens 1 one) × 10		
(5 hundreds 5 tens) × 10		
(2 thousands 7 tens) ÷ 10		
(4 ten thousands 8 hundreds) ÷ 10		

4. a. Emily collected $950 selling Girl Scout cookies all day Saturday. Emily's troop collected 10 times as much as she did. How much money did Emily's troop raise?

b. On Saturday, Emily made 10 times as much as on Monday. How much money did Emily collect on Monday?

unlabeled millions place value chart

This page intentionally left blank

Name ______________________________ Date ________________

1. Rewrite the following numbers including commas where appropriate:

a. 1234 ______________ b. 12345 ______________ c. 123456 ______________

d. 1234567 ______________ e. 12345678901 ______________

2. Solve each expression. Record your answer in standard form.

Expression	Standard Form
5 tens + 5 tens	
3 hundreds + 7 hundreds	
400 thousands + 600 thousands	
8 thousands + 4 thousands	

3. Represent each addend with place value disks in the place value chart. Show the composition of larger units from 10 smaller units. Write the sum in standard form.

a. 4 thousands + 11 hundreds = ______________________

millions	hundred thousands	ten thousands	thousands	hundreds	tens	ones

b. 24 ten thousands + 11 thousands = ____________________

millions	hundred thousands	ten thousands	thousands	hundreds	tens	ones

4. Use digits or disks on the place value chart to represent the following equations. Write the product in standard form.

a. 10 × 3 thousands = ____________________

How many thousands are in the answer? ____________

millions	hundred thousands	ten thousands	thousands	hundreds	tens	ones

b. (3 ten thousands 2 thousands) × 10 = ____________________

How many thousands are in the answer? ____________

millions	hundred thousands	ten thousands	thousands	hundreds	tens	ones

c. (32 thousands 1 hundred 4 ones) × 10 = ______________________

How many thousands are in your answer? ______________________

millions	hundred thousands	ten thousands	thousands	hundreds	tens	ones

5. Lee and Gary visited South Korea. They exchanged their dollars for South Korean bills. Lee received 15 ten thousand South Korean bills. Gary received 150 thousand bills. Use disks or numbers on a place value chart to compare Lee's and Gary's money.

Name ______________________________ Date ______________

1. Rewrite the following numbers including commas where appropriate:

a. 4321 ____________________ b. 54321 ____________________

c. 224466 ____________________ d. 2224466 ____________________

e. 10010011001 ____________________

2. Solve each expression. Record your answer in standard form.

Expression	Standard Form
4 tens + 6 tens	
8 hundreds + 2 hundreds	
5 thousands + 7 thousands	

3. Represent each addend with place value disks in the place value chart. Show the composition of larger units from 10 smaller units. Write the sum in standard form.

a. 2 thousands + 12 hundreds = ____________________

millions	hundred thousands	ten thousands	thousands	hundreds	tens	ones

b. 14 ten thousands + 12 thousands = ____________________

millions	hundred thousands	ten thousands	thousands	hundreds	tens	ones

4. Use digits or disks on the place value chart to represent the following equations. Write the product in standard form.

a. 10 × 5 thousands = ____________________

How many thousands are in the answer? ____________

millions	hundred thousands	ten thousands	thousands	hundreds	tens	ones

b. (4 ten thousands 4 thousands) × 10 = ____________________

How many thousands are in the answer? ____________

millions	hundred thousands	ten thousands	thousands	hundreds	tens	ones

c. (27 thousands 3 hundreds 5 ones) × 10 = ____________________

How many thousands are in your answer? ____________________

millions	hundred thousands	ten thousands	thousands	hundreds	tens	ones

5. A large grocery store received an order of 2 thousand apples. A neighboring school received an order of 20 boxes of apples with 100 apples in each. Use disks or disks on a place value chart to compare the number of apples received by the school and the number of apples received by the grocery store.

Name ______________________________ Date ________________

1. a. On the place value chart below, label the units, and represent the number 90,523.

   b. Write the number in word form.

   c. Write the number in expanded form.

2. a. On the place value chart below, label the units, and represent the number 905,203.

   b. Write the number in word form.

   c. Write the number in expanded form.

G4-M1-SE-B1-1.3.1-1.2016

3. Complete the following chart:

Standard Form	Word Form	Expanded Form
	two thousand, four hundred eighty	
		20,000 + 400 + 80 + 2
	sixty-four thousand, one hundred six	
604,016		
960,060		

4. Black rhinos are endangered, with only 4,400 left in the world. Timothy read that number as "four thousand, four hundred." His father read the number as "44 hundred." Who read the number correctly? Use pictures, numbers, or words to explain your answer.

Name ______________________________ Date ______________

1. a. On the place value chart below, label the units, and represent the number 50,679.

b. Write the number in word form.

c. Write the number in expanded form.

2. a. On the place value chart below, label the units, and represent the number 506,709.

b. Write the number in word form.

c. Write the number in expanded form.

3. Complete the following chart:

Standard Form	Word Form	Expanded Form
	five thousand, three hundred seventy	
		50,000 + 300 + 70 + 2
	thirty-nine thousand, seven hundred one	
309,017		
770,070		

4. Use pictures, numbers, and words to explain another way to say sixty-five hundred.

Name ______________________________ Date ______________

1. Label the units in the place value chart. Draw place value disks to represent each number in the place value chart. Use <, >, or = to compare the two numbers. Write the correct symbol in the circle.

   a. 600,015 ◯ 60,015

   b. 409,004 ◯ 440,002

2. Compare the two numbers by using the symbols <, >, and =. Write the correct symbol in the circle.

   a. 342,001 ◯ 94,981

   b. 500,000 + 80,000 + 9,000 + 100 ◯ five hundred eight thousand, nine hundred one

c. 9 hundred thousands 8 thousands 9 hundreds 3 tens ◯ 908,930

d. 9 hundreds 5 ten thousands 9 ones ◯ 6 ten thousands 5 hundreds 9 ones

3. Use the information in the chart below to list the height in feet of each mountain from least to greatest. Then, name the mountain that has the lowest elevation in feet.

Name of Mountain	Elevation in Feet (ft)
Allen Mountain	4,340 ft
Mount Marcy	5,344 ft
Mount Haystack	4,960 ft
Slide Mountain	4,240 ft

4. Arrange these numbers from least to greatest: 8,002 2,080 820 2,008 8,200

5. Arrange these numbers from greatest to least: 728,000 708,200 720,800 87,300

6. One astronomical unit, or 1 AU, is the approximate distance from Earth to the sun. The following are the approximate distances from Earth to nearby stars given in AUs:

   Alpha Centauri is 275,725 AUs from Earth.
   Proxima Centauri is 268,269 AUs from Earth.
   Epsilon Eridani is 665,282 AUs from Earth.
   Barnard's Star is 377,098 AUs from Earth.
   Sirius is 542,774 AUs from Earth.

   List the names of the stars and their distances in AUs in order from closest to farthest from Earth.

Name ______________________________ Date ________________

1. Label the units in the place value chart. Draw place value disks to represent each number in the place value chart. Use <, >, or = to compare the two numbers. Write the correct symbol in the circle.

a. 909,013 ◯ 90,013

b. 210,005 ◯ 220,005

2. Compare the two numbers by using the symbols <, >, and =. Write the correct symbol in the circle.

   a. 501,107 ◯ 89,171

   b. 300,000 + 50,000 + 1,000 + 800 ◯ six hundred five thousand, nine hundred eight

   c. 3 hundred thousands 3 thousands 8 hundreds 4 tens ◯ 303,840

   d. 5 hundreds 6 ten thousands 2 ones ◯ 3 ten thousands 5 hundreds 1 one

3. Use the information in the chart below to list the height, in feet, of each skyscraper from shortest to tallest. Then, name the tallest skyscraper.

Name of Skyscraper	Height of Skyscraper (ft)
Willis Tower	1,450 ft
One World Trade Center	1,776 ft
Taipei 101	1,670 ft
Petronas Towers	1,483 ft

G4-M1-SE-B1-1.3.1-1.2016

4. Arrange these numbers from least to greatest: 7,550 5,070 750 5,007 7,505

5. Arrange these numbers from greatest to least: 426,000 406,200 640,020 46,600

6. The areas of the 50 states can be measured in square miles.

   California is 158,648 square miles. Nevada is 110,567 square miles. Arizona is 114,007 square miles. Texas is 266,874 square miles. Montana is 147,047 square miles, and Alaska is 587,878 square miles.

   Arrange the states in order from least area to greatest area.

unlabeled hundred thousands place value chart

This page intentionally left blank

Name ______________________________________ Date ____________________

1. Label the place value chart. Use place value disks to find the sum or difference. Write the answer in standard form on the line.

   a. 10,000 more than six hundred five thousand, four hundred seventy-two is ____________________.

   b. 100 thousand less than 400,000 + 80,000 + 1,000 + 30 + 6 is ____________________.

   c. 230,070 is ______________________________________ than 130,070.

2. Lucy plays an online math game. She scored 100,000 more points on Level 2 than on Level 3. If she scored 349,867 points on Level 2, what was her score on Level 3? Use pictures, words, or numbers to explain your thinking.

3. Fill in the blank for each equation.

a. 10,000 + 40,060 = ____________

b. 21,195 – 10,000 = ____________

c. 999,000 + 1,000 = ____________

d. 129,231 – 100,000 = ____________

e. 122,000 = 22,000 + ____________

f. 38,018 = 39,018 – ____________

4. Fill in the empty boxes to complete the patterns.

a.

150,010		170,010		190,010	

Explain in pictures, numbers, or words how you found your answers.

b.

	898,756	798,756			498,756

Explain in pictures, numbers, or words how you found your answers.

c.

744,369	743,369		741,369		

Explain in pictures, numbers, or words how you found your answers.

d.

	118,910			88,910	78,910

Explain in pictures, numbers, or words how you found your answers.

Name ______________________________ Date ______________

1. Label the place value chart. Use place value disks to find the sum or difference. Write the answer in standard form on the line.

   a. 100,000 less than five hundred sixty thousand, three hundred thirteen is ____________.

   b. Ten thousand more than 300,000 + 90,000 + 5,000 + 40 is ________________.

   c. 447,077 is ______________________________ than 347,077.

2. Fill in the blank for each equation:

   a. 100,000 + 76,960 = ____________

   b. 13,097 – 1,000 = ____________

   c. 849,000 – 10,000 = ____________

   d. 442,210 + 10,000 = ____________

   e. 172,090 = 171,090 + ____________

   f. 854,121 = 954,121 – ____________

3. Fill in the empty boxes to complete the patterns.

a.

145,555		147,555		149,555	

Explain in pictures, numbers, or words how you found your answers.

b.

	764,321	774,321			804,321

Explain in pictures, numbers, or words how you found your answers.

c.

125,876	225,876		425,876		

Explain in pictures, numbers, or words how you found your answers.

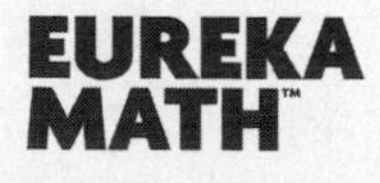

d.

	254,445			224,445	214,445

Explain in pictures, numbers, or words how you found your answers.

4. In 2012, Charlie earned an annual salary of $54,098. At the beginning of 2013, Charlie's annual salary was raised by $10,000. How much money will Charlie earn in 2013? Use pictures, words, or numbers to explain your thinking.

Name ____________________________ Date ________________

1. Round to the nearest thousand. Use the number line to model your thinking.

a. 6,700 ≈ ________________

b. 9,340 ≈ ________________

c. 16,401 ≈ ________________

d. 39,545 ≈ ________________

e. 399,499 ≈ ________________

f. 840,007 ≈ ________________

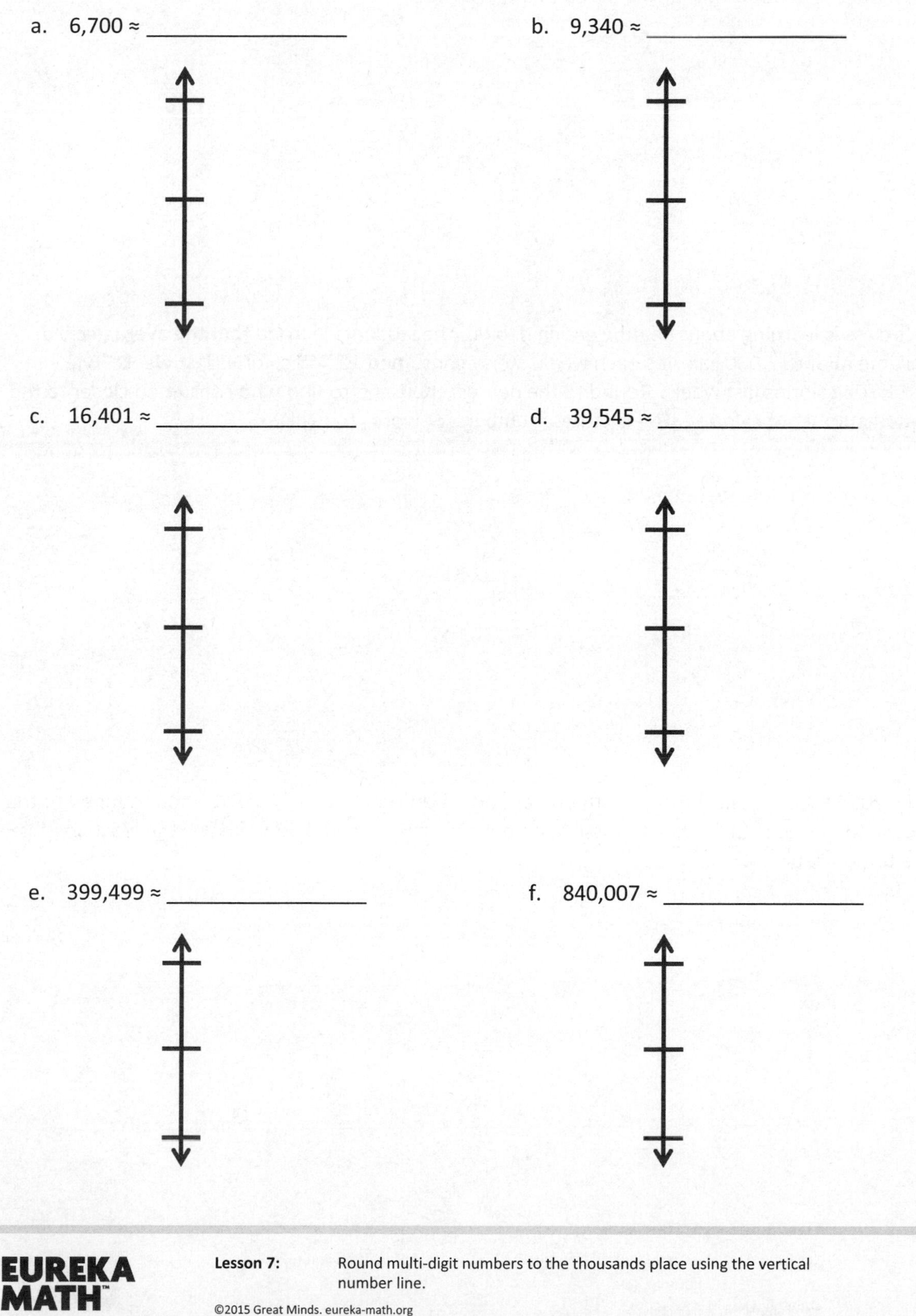

2. A pilot wanted to know about how many kilometers he flew on his last 3 flights. From NYC to London, he flew 5,572 km. Then, from London to Beijing, he flew 8,147 km. Finally, he flew 10,996 km from Beijing back to NYC. Round each number to the nearest thousand, and then find the sum of the rounded numbers to estimate about how many kilometers the pilot flew.

3. Mrs. Smith's class is learning about healthy eating habits. The students learned that the average child should consume about 12,000 calories each week. Kerry consumed 12,748 calories last week. Tyler consumed 11,702 calories last week. Round to the nearest thousand to find who consumed closer to the recommended number of calories. Use pictures, numbers, or words to explain.

4. For the 2013-2014 school year, the cost of tuition at Cornell University was $43,000 when rounded to the nearest thousand. What is the greatest possible amount the tuition could be? What is the least possible amount the tuition could be?

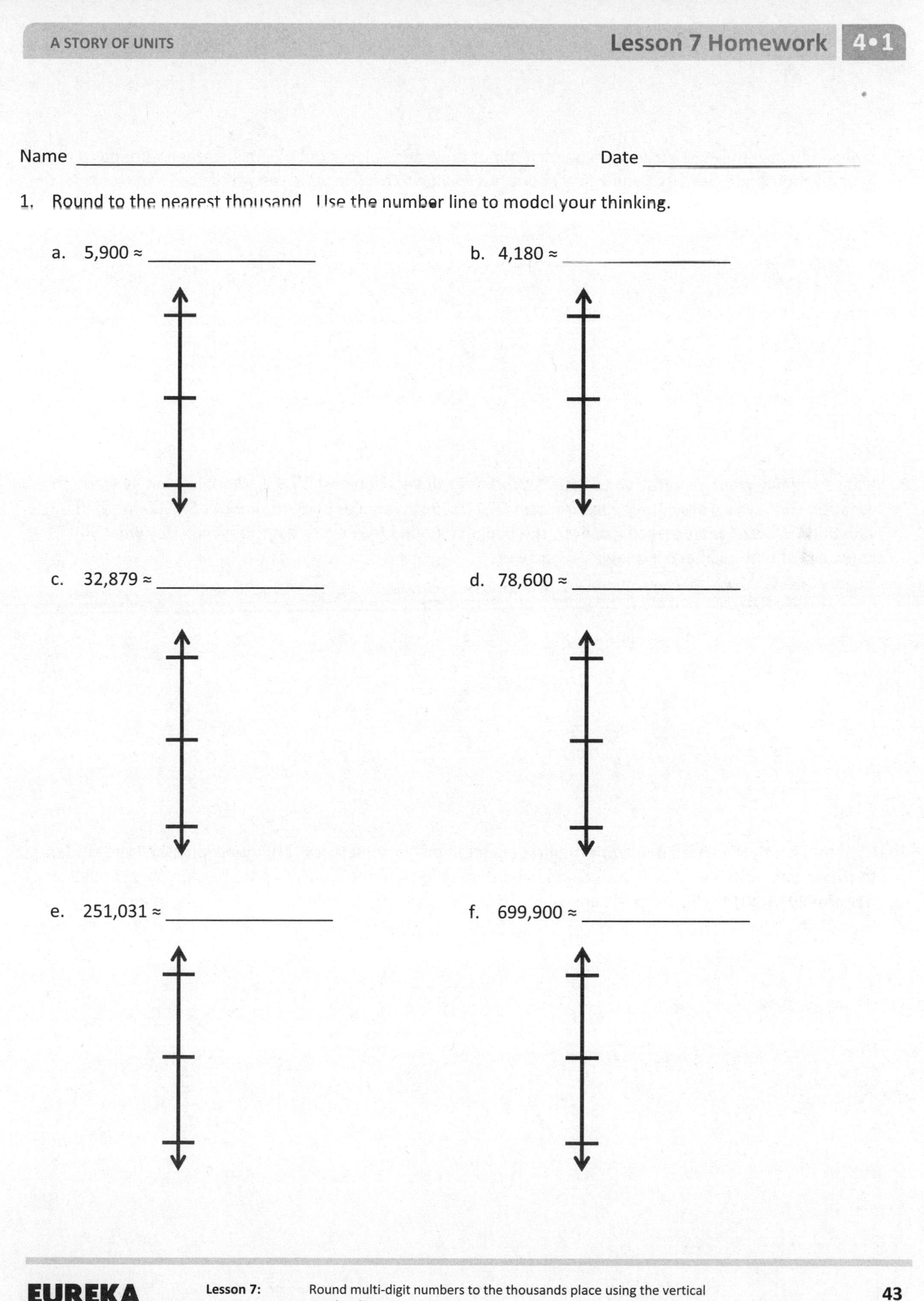

Name ______________________________ Date ____________

1. Round to the nearest thousand. Use the number line to model your thinking.

a. 5,900 ≈ ______________

b. 4,180 ≈ ______________

c. 32,879 ≈ ______________

d. 78,600 ≈ ______________

e. 251,031 ≈ ______________

f. 699,900 ≈ ______________

2. Steven put together 981 pieces of a puzzle. About how many pieces did he put together? Round to the nearest thousand. Use what you know about place value to explain your answer.

3. Louise's family went on vacation to Disney World. Their vacation cost $5,990. Sophia's family went on vacation to Niagara Falls. Their vacation cost $4,720. Both families budgeted about $5,000 for their vacation. Whose family stayed closer to the budget? Round to the nearest thousand. Use what you know about place value to explain your answer.

4. Marsha's brother wanted help with the first question on his homework. The question asked the students to round 128,902 to the nearest thousand and then to explain the answer. Marsha's brother thought that the answer was 128,000. Was his answer correct? How do you know? Use pictures, numbers, or words to explain.

Name ______________________________ Date ______________

Complete each statement by rounding the number to the given place value. Use the number line to show your work.

1. a. 53,000 rounded to the nearest ten thousand is ______________.

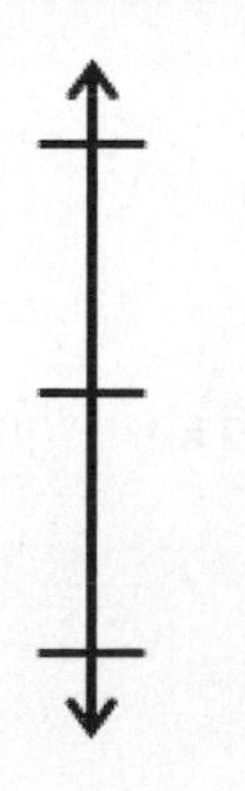

2. a. 240,000 rounded to the nearest hundred thousand is ______________.

b. 42,708 rounded to the nearest ten thousand is ______________.

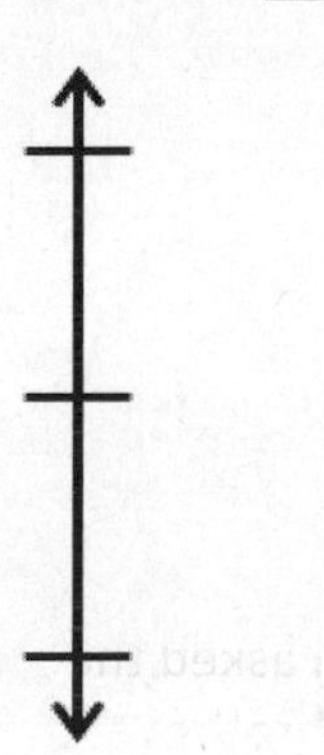

b. 449,019 rounded to the nearest hundred thousand is ______________.

c. 406,823 rounded to the nearest ten thousand is ______________.

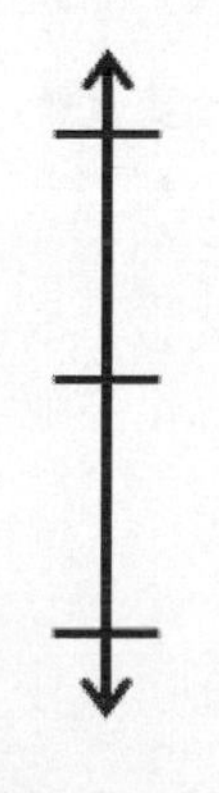

c. 964,103 rounded to the nearest hundred thousand is ______________.

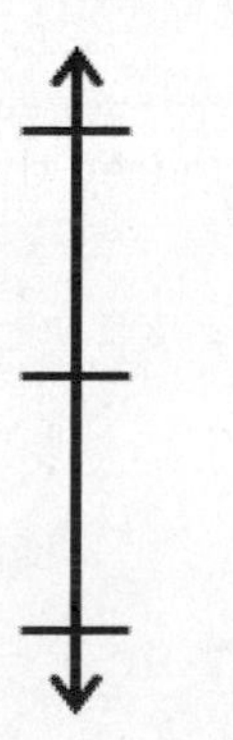

3. 975,462 songs were downloaded in one day. Round this number to the nearest hundred thousand to estimate how many songs were downloaded in one day. Use a number line to show your work.

4. This number was rounded to the nearest ten thousand. List the possible digits that could go in the thousands place to make this statement correct. Use a number line to show your work.

$$13\_,644 \approx 130,000$$

5. Estimate the difference by rounding each number to the given place value.

$$712,350 - 342,802$$

a. Round to the nearest ten thousands.

b. Round to the nearest hundred thousands.

Name ________________________ Date ____________

Complete each statement by rounding the number to the given place value. Use the number line to show your work.

1. a. 67,000 rounded to the nearest ten thousand is ______________.

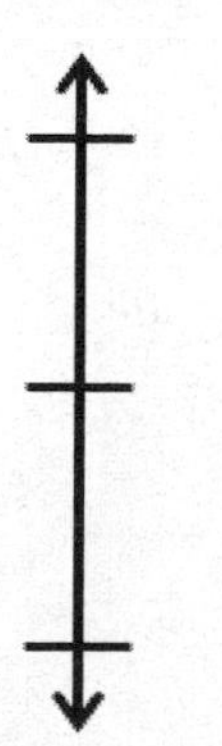

   b. 51,988 rounded to the nearest ten thousand is ______________.

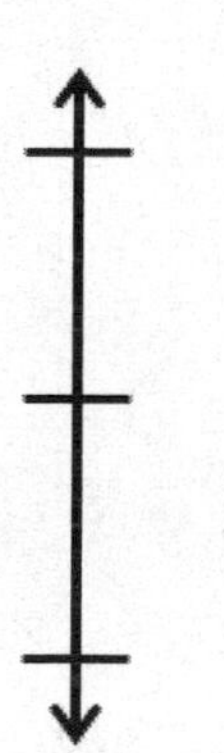

   c. 105,159 rounded to the nearest ten thousand is ______________.

2. a. 867,000 rounded to the nearest hundred thousand is ______________.

   b. 767,074 rounded to the nearest hundred thousand is ______________.

   c. 629,999 rounded to the nearest hundred thousand is ______________.

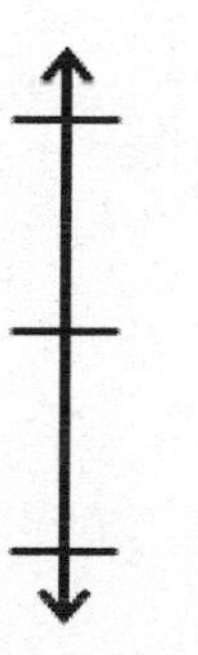

3. 491,852 people went to the water park in the month of July. Round this number to the nearest hundred thousand to estimate how many people went to the park. Use a number line to show your work.

4. This number was rounded to the nearest hundred thousand. List the possible digits that could go in the ten thousands place to make this statement correct. Use a number line to show your work.

$$1\_9,644 \approx 100,000$$

5. Estimate the sum by rounding each number to the given place value.

$$164,215 + 216,088$$

a. Round to the nearest ten thousand.

b. Round to the nearest hundred thousand.

G4-M1-SE-B1-1.3.1-1.2016

Name ______________________________ Date ______________

1. Round to the nearest thousand.

   a. 5,300 ≈ ____________________ b. 4,589 ≈ ____________________

   c. 42,099 ≈ ____________________ d. 801,504 ≈ ____________________

   e. Explain how you found your answer for Part (d).

2. Round to the nearest ten thousand.

   a. 26,000 ≈ ____________________ b. 34,920 ≈ ____________________

   c. 789,091 ≈ ____________________ d. 706,286 ≈ ____________________

   e. Explain why two problems have the same answer. Write another number that has the same answer when rounded to the nearest ten thousand.

3. Round to the nearest hundred thousand.

   a. 840,000 ≈ ____________________ b. 850,471 ≈ ____________________

   c. 761,004 ≈ ____________________ d. 991,965 ≈ ____________________

   e. Explain why two problems have the same answer. Write another number that has the same answer when rounded to the nearest hundred thousand.

G4-M1-SE-B1-1.3.1-1.2016

4. Solve the following problems using pictures, numbers, or words.

   a. The 2012 Super Bowl had an attendance of just 68,658 people. If the headline in the newspaper the next day read, "About 70,000 People Attend Super Bowl," how did the newspaper round to estimate the total number of people in attendance?

   b. The 2011 Super Bowl had an attendance of 103,219 people. If the headline in the newspaper the next day read, "About 200,000 People Attend Super Bowl," is the newspaper's estimate reasonable? Use rounding to explain your answer.

   c. According to the problems above, about how many more people attended the Super Bowl in 2011 than in 2012? Round each number to the largest place value before giving the estimated answer.

Name ______________________________ Date ________________

1. Round to the nearest thousand.

 a. 6,842 ≈ ________________ b. 2,722 ≈ ________________

 c. 16,051 ≈ ________________ d. 706,421 ≈ ________________

 e. Explain how you found your answer for Part (d).

2. Round to the nearest ten thousand.

 a. 88,999 ≈ ________________ b. 85,001 ≈ ________________

 c. 789,091 ≈ ________________ d. 905,154 ≈ ________________

 e. Explain why two problems have the same answer. Write another number that has the same answer when rounded to the nearest ten thousand.

3. Round to the nearest hundred thousand.

 a. 89,659 ≈ ________________ b. 751,447 ≈ ________________

 c. 617,889 ≈ ________________ d. 817,245 ≈ ________________

 e. Explain why two problems have the same answer. Write another number that has the same answer when rounded to the nearest hundred thousand.

4. Solve the following problems using pictures, numbers, or words.

   a. At President Obama's inauguration in 2013, the newspaper headlines stated there were about 800,000 people in attendance. If the newspaper rounded to the nearest hundred thousand, what is the largest number and smallest number of people who could have been there?

   b. At President Bush's inauguration in 2005, the newspaper headlines stated there were about 400,000 people in attendance. If the newspaper rounded to the nearest ten thousand, what is the largest number and smallest number of people who could have been there?

   c. At President Lincoln's inauguration in 1861, the newspaper headlines stated there were about 30,000 people in attendance. If the newspaper rounded to the nearest thousand, what is the largest number and smallest number of people who could have been there?

Name ______________________________ Date ______________

1. Round 543,982 to the nearest

   a. thousand: ______________________.

   b. ten thousand: ______________________.

   c. hundred thousand: ______________________.

2. Complete each statement by rounding the number to the given place value.

   a. 2,841 rounded to the nearest hundred is ______________________.

   b. 32,851 rounded to the nearest hundred is ______________________.

   c. 132,891 rounded to the nearest hundred is ______________________.

   d. 6,299 rounded to the nearest thousand is ______________________.

   e. 36,599 rounded to the nearest thousand is ______________________.

   f. 100,699 rounded to the nearest thousand is ______________________.

   g. 40,984 rounded to the nearest ten thousand is ______________________.

   h. 54,984 rounded to the nearest ten thousand is ______________________.

   i. 997,010 rounded to the nearest ten thousand is ______________________.

   j. 360,034 rounded to the nearest hundred thousand is ______________________.

   k. 436,709 rounded to the nearest hundred thousand is ______________________.

   l. 852,442 rounded to the nearest hundred thousand is ______________________.

3. Empire Elementary School needs to purchase water bottles for field day. There are 2,142 students. Principal Vadar rounded to the nearest hundred to estimate how many water bottles to order. Will there be enough water bottles for everyone? Explain.

4. Opening day at the New York State Fair in 2012 had an attendance of 46,753. Decide which place value to round 46,753 to if you were writing a newspaper article. Round the number and explain why it is an appropriate unit to round the attendance to.

5. A jet airplane holds about 65,000 gallons of gas. It uses about 7,460 gallons when flying between New York City and Los Angeles. Round each number to the largest place value. Then, find about how many trips the plane can take between cities before running out of fuel.

Name ______________________________ Date ______________

1. Round 845,001 to the nearest

   a. thousand: ______________________.

   b. ten thousand: ______________________.

   c. hundred thousand: ______________________.

2. Complete each statement by rounding the number to the given place value.

   a. 783 rounded to the nearest hundred is ______________________.

   b. 12,781 rounded to the nearest hundred is ______________________.

   c. 951,194 rounded to the nearest hundred is ______________________.

   d. 1,258 rounded to the nearest thousand is ______________________.

   e. 65,124 rounded to the nearest thousand is ______________________.

   f. 99,451 rounded to the nearest thousand is ______________________.

   g. 60,488 rounded to the nearest ten thousand is ______________________.

   h. 80,801 rounded to the nearest ten thousand is ______________________.

   i. 897,100 rounded to the nearest ten thousand is ______________________.

   j. 880,005 rounded to the nearest hundred thousand is ______________________.

   k. 545,999 rounded to the nearest hundred thousand is ______________________.

   l. 689,114 rounded to the nearest hundred thousand is ______________________.

3. Solve the following problems using pictures, numbers, or words.

   a. In the 2011 New York City Marathon, 29,867 men finished the race, and 16,928 women finished the race. Each finisher was given a t-shirt. About how many men's shirts were given away? About how many women's shirts were given away? Explain how you found your answers.

   b. In the 2010 New York City Marathon, 42,429 people finished the race and received a medal. Before the race, the medals had to be ordered. If you were the person in charge of ordering the medals and estimated how many to order by rounding, would you have ordered enough medals? Explain your thinking.

   c. In 2010, 28,357 of the finishers were men, and 14,072 of the finishers were women. About how many more men finished the race than women? To determine your answer, did you round to the nearest ten thousand or thousand? Explain.

Name ______________________________ Date ____________________

1. Solve the addition problems below using the standard algorithm.

a.
```
  6, 3 1 1
+     2 6 8
```

b.
```
  6, 3 1 1
+ 1, 2 6 8
```

c.
```
  6, 3 1 4
+ 1, 2 6 8
```

d.
```
  6, 3 1 4
+ 2, 4 9 3
```

e.
```
  8, 3 1 4
+ 2, 4 9 3
```

f.
```
  1 2, 3 7 8
+     5, 4 6 3
```

g.
```
  5 2, 0 9 8
+     6, 0 4 8
```

h.
```
  3 4, 6 9 8
+ 7 1, 8 4 0
```

i.
```
  5 4 4, 8 1 1
+ 3 5 6, 4 4 5
```

j. 527 + 275 + 752

k. 38,193 + 6,376 + 241,457

Draw a tape diagram to represent each problem. Use numbers to solve, and write your answer as a statement.

2. In September, Liberty Elementary School collected 32,537 cans for a fundraiser. In October, they collected 207,492 cans. How many cans were collected during September and October?

3. A baseball stadium sold some burgers. 2,806 were cheeseburgers. 1,679 burgers didn't have cheese. How many burgers did they sell in all?

4. On Saturday night, 23,748 people attended the concert. On Sunday, 7,570 more people attended the concert than on Saturday. How many people attended the concert on Sunday?

G4-M1-SE-B1-1.3.1-1.2016

Name ______________________________ Date ______________

1. Solve the addition problems below using the standard algorithm.

a.
```
   7,909
+  1,044
```

b.
```
  27,909
+  9,740
```

c.
```
  827,909
+  42,989
```

d.
```
  289,205
+  11,845
```

e.
```
  547,982
+ 114,849
```

f.
```
  258,983
+ 121,897
```

g.
```
  83,906
+ 35,808
```

h.
```
  289,999
+  91,849
```

i.
```
  754,900
+ 245,100
```

Draw a tape diagram to represent each problem. Use numbers to solve, and write your answer as a statement.

2. At the zoo, Brooke learned that one of the rhinos weighs 4,897 pounds, one of the giraffes weighs 2,667 pounds, one of the African elephants weighs 12,456 pounds, and one of the Komodo dragons weighs 123 pounds.

   a. What is the combined weight of the zoo's African elephant and the giraffe?

   b. What is the combined weight of the zoo's African elephant and the rhino?

   c. What is the combined weight of the zoo's African elephant, the rhino, and the giraffe?

   d. What is the combined weight of the zoo's Komodo dragon and the rhino?

millions	hundred thousands	ten thousands	thousands	hundreds	tens	ones

millions place value chart

This page intentionally left blank

Name ______________________________ Date ______________

Estimate and then solve each problem. Model the problem with a tape diagram. Explain if your answer is reasonable.

1. For the bake sale, Connie baked 144 cookies. Esther baked 49 more cookies than Connie.

   a. About how many cookies did Connie and Esther bake? Estimate by rounding each number to the nearest ten before adding.

   b. Exactly how many cookies did Connie and Esther bake?

   c. Is your answer reasonable? Compare your estimate from (a) to your answer from (b). Write a sentence to explain your reasoning.

2. Raffle tickets were sold for a school fundraiser to parents, teachers, and students. 563 tickets were sold to teachers. 888 more tickets were sold to students than to teachers. 904 tickets were sold to parents.

   a. About how many tickets were sold to parents, teachers, and students? Round each number to the nearest hundred to find your estimate.

   b. Exactly how many tickets were sold to parents, teachers, and students?

   c. Assess the reasonableness of your answer in (b). Use your estimate from (a) to explain.

3. From 2010 to 2011, the population of Queens increased by 16,075. Brooklyn's population increased by 11,870 more than the population increase of Queens.

   a. Estimate the total combined population increase of Queens and Brooklyn from 2010 to 2011. (Round the addends to estimate.)

   b. Find the actual total combined population increase of Queens and Brooklyn from 2010 to 2011.

   c. Assess the reasonableness of your answer in (b). Use your estimate from (a) to explain.

4. During National Recycling Month, Mr. Yardley's class spent 4 weeks collecting empty cans to recycle.

Week	Number of Cans Collected
1	10,827
2	
3	10,522
4	20,011

a. During Week 2, the class collected 1,256 more cans than they did during Week 1. Find the total number of cans Mr. Yardley's class collected in 4 weeks.

b. Assess the reasonableness of your answer in (a) by estimating the total number of cans collected.

Name ______________________________ Date ________________

Estimate and then solve each problem. Model the problem with a tape diagram. Explain if your answer is reasonable.

1. There were 3,905 more hits on the school's website in January than February. February had 9,854 hits. How many hits did the school's website have during both months?

   a. About how many hits did the website have during January and February?

   b. Exactly how many hits did the website have during January and February?

   c. Is your answer reasonable? Compare your estimate from (a) to your answer from (b). Write a sentence to explain your reasoning.

2. On Sunday, 77,098 fans attended a New York Jets game. The same day, 3,397 more fans attended a New York Giants game than attended the Jets game. Altogether, how many fans attended the games?

   a. What was the actual number of fans who attended the games?

   b. Is your answer reasonable? Round each number to the nearest thousand to find an estimate of how many fans attended the games.

3. Last year on Ted's farm, his four cows produced the following number of liters of milk:

Cow	Liters of Milk Produced
Daisy	5,098
Betsy	
Mary	9,980
Buttercup	7,087

a. Betsy produced 986 more liters of milk than Buttercup. How many liters of milk did all 4 cows produce?

b. Is your answer reasonable? Explain.

This page intentionally left blank

Name ______________________________ Date ______________

1. Use the standard algorithm to solve the following subtraction problems.

a. 7, 5 2 5
 − 3, 5 0 2

b. 1 7, 5 2 5
 − 1 3, 5 0 2

c. 6, 6 2 5
 − 4, 4 1 7

d. 4, 6 2 5
 − 4 3 5

e. 6, 5 0 0
 − 4 7 0

f. 6, 0 2 5
 − 3, 5 0 2

g. 2 3, 6 4 0
 − 1 4, 6 3 0

h. 4 3 1, 9 2 5
 − 2 0 4, 8 1 5

i. 2 1 9, 9 2 5
 − 1 2 1, 7 0 5

Draw a tape diagram to represent each problem. Use numbers to solve, and write your answer as a statement. Check your answers.

2. What number must be added to 13,875 to result in a sum of 25,884?

3. Artist Michelangelo was born on March 6, 1475. Author Mem Fox was born on March 6, 1946. How many years after Michelangelo was born was Fox born?

4. During the month of March, 68,025 pounds of king crab were caught. If 15,614 pounds were caught in the first week of March, how many pounds were caught in the rest of the month?

5. James bought a used car. After driving exactly 9,050 miles, the odometer read 118,064 miles. What was the odometer reading when James bought the car?

Name ______________________________ Date ______________

1. Use the standard algorithm to solve the following subtraction problems.

a.
```
  2,431
-   341
```

b.
```
  422,431
-  14,321
```

c.
```
  422,431
-  92,420
```

d.
```
  422,431
- 392,420
```

e.
```
  982,430
-  92,300
```

f.
```
  243,089
- 137,079
```

g. 2,431 – 920 =

h. 892,431 – 520,800 =

2. What number must be added to 14,056 to result in a sum of 38,773?

Draw a tape diagram to model each problem. Use numbers to solve, and write your answers as a statement. Check your answers.

3. An elementary school collected 1,705 bottles for a recycling program. A high school also collected some bottles. Both schools collected 3,627 bottles combined. How many bottles did the high school collect?

4. A computer shop sold $356,291 worth of computers and accessories. It sold $43,720 worth of accessories. How much did the computer shop sell in computers?

5. The population of a city is 538,381. In that population, 148,170 are children.

   a. How many adults live in the city?

   b. 186,101 of the adults are males. How many adults are female?

G4-M1-SE-B1-1.3.1-1.2016

This page intentionally left blank

Name ______________________________ Date ______________

1. Use the standard algorithm to solve the following subtraction problems.

a.
```
  2,460
- 1,370
```

b.
```
  2,460
- 1,470
```

c.
```
  97,684
- 49,700
```

d.
```
  2,460
- 1,472
```

e.
```
  124,306
-  31,117
```

f.
```
  97,684
-  4,705
```

g.
```
  124,006
- 121,117
```

h.
```
  97,684
- 47,705
```

i.
```
  124,060
-  31,117
```

Draw a tape diagram to represent each problem. Use numbers to solve, and write your answer as a statement. Check your answers.

2. There are 86,400 seconds in one day. If Mr. Liegel is at work for 28,800 seconds a day, how many seconds a day is he away from work?

3. A newspaper company delivered 240,900 newspapers before 6 a.m. on Sunday. There were a total of 525,600 newspapers to deliver. How many more newspapers needed to be delivered on Sunday?

4. A theater holds a total of 2,013 chairs. 197 chairs are in the VIP section. How many chairs are not in the VIP section?

5. Chuck's mom spent $19,155 on a new car. She had $30,064 in her bank account. How much money does Chuck's mom have after buying the car?

Name ____________________ Date ____________

1. Use the standard algorithm to solve the following subtraction problems.

a. $\begin{array}{r} 71,989 \\ -\ 21,492 \\ \hline \end{array}$

b. $\begin{array}{r} 371,989 \\ -\ 96,492 \\ \hline \end{array}$

c. $\begin{array}{r} 371,089 \\ -\ 25,192 \\ \hline \end{array}$

d. $\begin{array}{r} 879,989 \\ -\ 721,492 \\ \hline \end{array}$

e. $\begin{array}{r} 879,009 \\ -\ 788,492 \\ \hline \end{array}$

f. $\begin{array}{r} 879,989 \\ -\ 21,070 \\ \hline \end{array}$

g. $\begin{array}{r} 879,000 \\ -\ 21,989 \\ \hline \end{array}$

h. $\begin{array}{r} 279,389 \\ -\ 191,492 \\ \hline \end{array}$

i. $\begin{array}{r} 500,989 \\ -\ 242,000 \\ \hline \end{array}$

Draw a tape diagram to represent each problem. Use numbers to solve, and write your answer as a statement. Check your answers.

2. Jason ordered 239,021 pounds of flour to be used in his 25 bakeries. The company delivering the flour showed up with 451,202 pounds. How many extra pounds of flour were delivered?

3. In May, the New York Public Library had 124,061 books checked out. Of those books, 31,117 were mystery books. How many of the books checked out were not mystery books?

4. A Class A dump truck can haul 239,000 pounds of dirt. A Class C dump truck can haul 600,200 pounds of dirt. How many more pounds can a Class C truck haul than a Class A truck?

Name ______________________________ Date ____________

1. Use the standard subtraction algorithm to solve the problems below.

a.
```
  1 0 1, 6 6 0
−     9 1, 6 8 0
```

b.
```
  1 0 1, 6 6 0
−       9, 9 8 0
```

c.
```
  2 4 2, 5 6 1
−     4 4, 7 0 2
```

d.
```
  2 4 2, 5 6 1
−     7 4, 9 8 7
```

e.
```
  1, 0 0 0, 0 0 0
−       5 9 2, 0 0 0
```

f.
```
  1, 0 0 0, 0 0 0
−       5 9 2, 5 0 0
```

g.
```
  6 0 0, 6 5 8
− 5 9 2, 5 6 9
```

h.
```
  6 0 0, 0 0 0
− 5 9 2, 5 6 9
```

Use tape diagrams and the standard algorithm to solve the problems below. Check your answers.

2. David is flying from Hong Kong to Buenos Aires. The total flight distance is 11,472 miles. If the plane has 7,793 miles left to travel, how far has it already traveled?

3. Tank A holds 678,500 gallons of water. Tank B holds 905,867 gallons of water. How much less water does Tank A hold than Tank B?

4. Mark had $25,081 in his bank account on Thursday. On Friday, he added his paycheck to the bank account, and he then had $26,010 in the account. What was the amount of Mark's paycheck?

Name ______________________________ Date ________________

1 . Use the standard subtraction algorithm to solve the problems below

a. 9,656 − 838

b. 59,656 − 5,880

c. 759,656 − 579,989

d. 294,150 − 166,370

e. 294,150 − 239,089

f. 294,150 − 96,400

g. 800,500 − 79,989

h. 800,500 − 45,500

i. 800,500 − 276,664

Use tape diagrams and the standard algorithm to solve the problems below. Check your answers.

2. A fishing boat was out to sea for 6 months and traveled a total of 8,578 miles. In the first month, the boat traveled 659 miles. How many miles did the fishing boat travel during the remaining 5 months?

3. A national monument had 160,747 visitors during the first week of September. A total of 759,656 people visited the monument in September. How many people visited the monument in September after the first week?

4. Shadow Software Company earned a total of $800,000 selling programs during the year 2012. $125,300 of that amount was used to pay expenses of the company. How much profit did Shadow Software Company make in the year 2012?

5. At the local aquarium, Bubba the Seal ate 25,634 grams of fish during the week. If, on the first day of the week, he ate 6,987 grams of fish, how many grams of fish did he eat during the remainder of the week?

Name ______________________________ Date ______________

Estimate first, and then solve each problem. Model the problem with a tape diagram. Explain if your answer is reasonable.

1. On Monday, a farmer sold 25,196 pounds of potatoes. On Tuesday, he sold 18,023 pounds. On Wednesday, he sold some more potatoes. In all, he sold 62,409 pounds of potatoes.

   a. About how many pounds of potatoes did the farmer sell on Wednesday? Estimate by rounding each value to the nearest thousand, and then compute.

   b. Find the precise number of pounds of potatoes sold on Wednesday.

   c. Is your precise answer reasonable? Compare your estimate from (a) to your answer from (b). Write a sentence to explain your reasoning.

2. A gas station had two pumps. Pump A dispensed 241,752 gallons. Pump B dispensed 113,916 more gallons than Pump A.

   a. About how many gallons did both pumps dispense? Estimate by rounding each value to the nearest hundred thousand and then compute.

   b. Exactly how many gallons did both pumps dispense?

   c. Assess the reasonableness of your answer in (b). Use your estimate from (a) to explain.

3. Martin's car had 86,456 miles on it. Of that distance, Martin's wife drove 24,901 miles, and his son drove 7,997 miles. Martin drove the rest.

   a. About how many miles did Martin drive? Round each value to estimate.

   b. Exactly how many miles did Martin drive?

   c. Assess the reasonableness of your answer in (b). Use your estimate from (a) to explain.

4. A class read 3,452 pages the first week and 4,090 more pages in the second week than in the first week. How many pages had they read by the end of the second week? Is your answer reasonable? Explain how you know using estimation.

5. A cargo plane weighed 500,000 pounds. After the first load was taken off, the airplane weighed 437,981 pounds. Then 16,478 more pounds were taken off. What was the total number of pounds of cargo removed from the plane? Is your answer reasonable? Explain.

Name ______________________________________ Date ____________________

1. Zachary's final project for a college course took a semester to write and had 95,234 words. Zachary wrote 35,295 words the first month and 19,240 words the second month.

   a. Round each value to the nearest ten thousand to estimate how many words Zachary wrote during the remaining part of the semester.

   b. Find the exact number of words written during the remaining part of the semester.

   c. Use your answer from (a) to explain why your answer in (b) is reasonable.

2. During the first quarter of the year, 351,875 people downloaded an app for their smartphones. During the second quarter of the year, 101,949 fewer people downloaded the app than during the first quarter. How many downloads occurred during the two quarters of the year?

   a. Round each number to the nearest hundred thousand to estimate how many downloads occurred during the first two quarters of the year.

   b. Determine exactly how many downloads occurred during the first two quarters of the year.

   c. Determine if your answer is reasonable. Explain.

3. A local store was having a two-week Back to School sale. They started the sale with 36,390 notebooks. During the first week of the sale, 7,424 notebooks were sold. During the second week of the sale, 8,967 notebooks were sold. How many notebooks were left at the end of the two weeks? Is your answer reasonable?

This page intentionally left blank

Name ______________________________ Date ________________

Draw a tape diagram to represent each problem. Use numbers to solve, and write your answer as a statement.

1. Sean's school raised $32,587. Leslie's school raised $18,749. How much more money did Sean's school raise?

2. At a parade, 97,853 people sat in bleachers, and 388,547 people stood along the street. How many fewer people were in the bleachers than standing on the street?

3. A pair of hippos weighs 5,201 kilograms together. The female weighs 2,038 kilograms. How much more does the male weigh than the female?

4. A copper wire was 240 meters long. After 60 meters was cut off, it was double the length of a steel wire. How much longer was the copper wire than the steel wire at first?

G4-M1-SE-B1-1.3.1-1.2016

Name ____________________ Date ____________

Draw a tape diagram to represent each problem. Use numbers to solve, and write your answer as a statement.

1. Gavin has 1,094 toy building blocks. Avery only has 816 toy building blocks. How many more building blocks does Gavin have?

2. Container B holds 2,391 liters of water. Together, Container A and Container B hold 11,875 liters of water. How many more liters of water does Container A hold than Container B?

3. A piece of yellow yarn was 230 inches long. After 90 inches had been cut from it, the piece of yellow yarn was twice as long as a piece of blue yarn. At first, how much longer was the yellow yarn than the blue yarn?

Name ______________________________ Date ______________

Draw a tape diagram to represent each problem. Use numbers to solve, and write your answer as a statement.

1. In one year, the factory used 11,650 meters of cotton, 4,950 fewer meters of silk than cotton, and 3,500 fewer meters of wool than silk. How many meters in all were used of the three fabrics?

2. The shop sold 12,789 chocolate and 9,324 cookie dough cones. It sold 1,078 more peanut butter cones than cookie dough cones and 999 more vanilla cones than chocolate cones. What was the total number of ice cream cones sold?

3. In the first week of June, a restaurant sold 10,345 omelets. In the second week, 1,096 fewer omelets were sold than in the first week. In the third week, 2 thousand more omelets were sold than in the first week. In the fourth week, 2 thousand fewer omelets were sold than in the first week. How many omelets were sold in all in June?

Name ______________________________ Date ______________

Draw a tape diagram to represent each problem. Use numbers to solve, and write your answer as a statement.

1. There were 22,869 children, 49,563 men, and 2,872 more women than men at the fair. How many people were at the fair?

2. Number A is 4,676. Number B is 10,043 greater than A. Number C is 2,610 less than B. What is the total value of numbers A, B, and C?

3. A store sold a total of 21,650 balls. It sold 11,795 baseballs. It sold 4,150 fewer basketballs than baseballs. The rest of the balls sold were footballs. How many footballs did the store sell?

Name ______________________________ Date ______________

Using the diagrams below, create your own word problem. Solve for the value of the variable.

1.

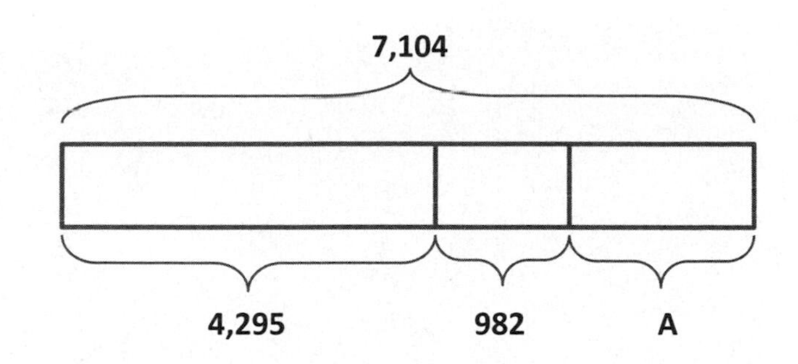

2.

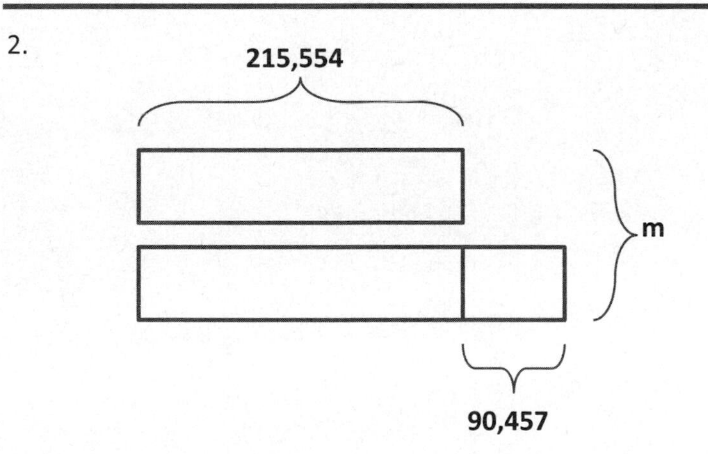

3.

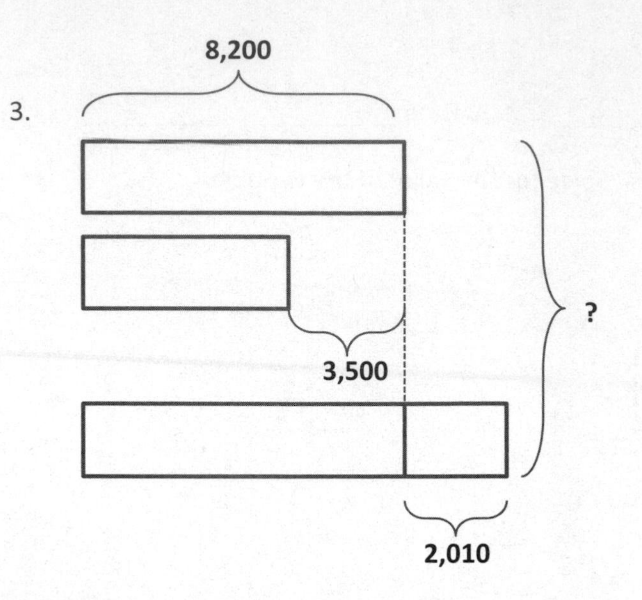

4. Draw a tape diagram to model the following equation. Create a word problem. Solve for the value of the variable.

$$26,854 = 17,729 + 3,731 + A$$

Name ______________________________ Date ________________

Using the diagrams below, create your own word problem. Solve for the value of the variable.

1. At the local botanical gardens, there are ____________ Redwoods and ____________ Cypress trees.

There are a total of ____________ Redwood, Cypress, and Dogwood trees.

How many ______________________________
____
______________________________?

Redwood | Cypress | Dogwood

6,294	3,849	A

12,115

---

2. There are 65,302 ______________________________
______________________________.

There are 37,436 fewer ______________________________
______________________________.

How many ______________________________
______________________________?

65,302

T

37,436

3. Use the following tape diagram to create a word problem. Solve for the value of the variable.

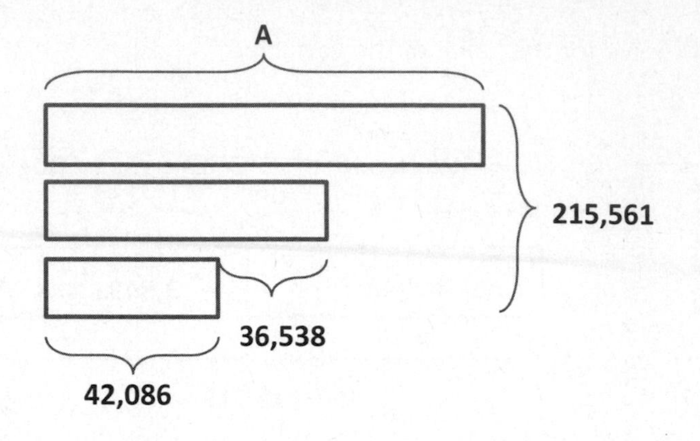

4. Draw a tape diagram to model the following equation. Create a word problem. Solve for the value of the variable.

$$27{,}894 + A + 6{,}892 = 40{,}392$$

Student Edition

Eureka Math
Grade 4
Module 2

Special thanks go to the Gordon A. Cain Center and to the Department of Mathematics at Louisiana State University for their support in the development of *Eureka Math*.

For a free *Eureka Math* Teacher Resource Pack, Parent Tip Sheets, and more please visit www.Eureka.tools

Published by the non-profit Great Minds

Printed in the U.S.A.
This book may be purchased from the publisher at eureka-math.org
10 9 8 7 6 5 4 3 2

ISBN 978-1-63255-303-4

Name ______________________________ Date ________________

1. Convert the measurements.

a. 1 km = ____________ m

b. 4 km = ____________ m

c. 7 km = ____________ m

d. ____________ km = 18,000 m

e. 1 m = ____________ cm

f. 3 m = ____________ cm

g. 80 m = ____________ cm

h. ____________ m = 12,000 cm

2. Convert the measurements.

a. 3 km 312 m = ____________ m

b. 13 km 27 m = ____________ m

c. 915 km 8 m = ____________ m

d. 3 m 56 cm = ____________ cm

e. 14 m 8 cm = ____________ cm

f. 120 m 46 cm = ____________ cm

3. Solve.

a. 4 km − 280 m

b. 1 m 15 cm − 34 cm

c. Express your answer in the smaller unit:
1 km 431 m + 13 km 169 m

d. Express your answer in the smaller unit:
231 m 31 cm − 14 m 48 cm

e. 67 km 230 m + 11 km 879 m

f. 67 km 230 m − 11 km 879 m

Use a tape diagram to model each problem. Solve using a simplifying strategy or an algorithm, and write your answer as a statement.

4. The length of Carter's driveway is 12 m 38 cm. His neighbor's driveway is 4 m 99 cm longer. How long is his neighbor's driveway?

5. Enya walked 2 km 309 m from school to the store. Then, she walked from the store to her home. If she walked a total of 5 km, how far was it from the store to her home?

6. Rachael has a rope 5 m 32 cm long that she cut into two pieces. One piece is 249 cm long. How many centimeters long is the other piece of rope?

7. Jason rode his bike 529 fewer meters than Allison. Jason rode 1 km 850 m. How many meters did Allison ride?

Name ____________________________ Date ______________

1. Find the equivalent measures.

a. 5 km = ____________ m

b. 13 km = ____________ m

c. ____________ km = 17,000 m

d. 60 km = ____________ m

e. 7 m = ____________ cm

f. 19 m = ____________ cm

g. ____________ m = 2,400 cm

h. 90 m = ____________ cm

2. Find the equivalent measures.

a. 7 km 123 m = ____________ m

b. 22 km 22 m = ____________ m

c. 875 km 4 m = ____________ m

d. 7 m 45 cm = ____________ cm

e. 67 m 7 cm = ____________ cm

f. 204 m 89 cm = ____________ cm

3. Solve.

a. 2 km 303 m – 556 m

b. 2 m – 54 cm

c. Express your answer in the smaller unit:
338 km 853 m + 62 km 71 m

d. Express your answer in the smaller unit:
800 m 35 cm – 154 m 49 cm

e. 701 km – 523 km 445 m

f. 231 km 811 m + 485 km 829 m

Use a tape diagram to model each problem. Solve using a simplifying strategy or an algorithm, and write your answer as a statement.

4. The length of Celia's garden is 15 m 24 cm. The length of her friend's garden is 2 m 98 cm more than Celia's. What is the length of her friend's garden?

5. Sylvia ran 3 km 290 m in the morning. Then, she ran some more in the evening. If she ran a total of 10 km, how far did Sylvia run in the evening?

6. Jenny's sprinting distance was 356 meters shorter than Tyler's. Tyler sprinted a distance of 1 km 3 m. How many meters did Jenny sprint?

7. The electrician had 7 m 23 cm of electrical wire. He used 551 cm for one wiring project. How many centimeters of wire does he have left?

Name ______________________________ Date ________________

1. Complete the conversion table.

Mass	
kg	**g**
1	1,000
3	
	4,000
17	
	20,000
300	

2. Convert the measurements.

a. 1 kg 500 g = ____________ g

b. 3 kg 715 g = ____________ g

c. 17 kg 84 g = ____________ g

d. 25 kg 9 g = ____________ g

e. _____ kg ______ g = 7,481 g

f. 210 kg 90 g = ____________ g

3. Solve.

a. 3,715 g − 1,500 g

b. 1 kg − 237 g

c. Express the answer in the smaller unit:
25 kg 9 g + 24 kg 991 g

d. Express the answer in the smaller unit:
27 kg 650 g − 20 kg 990 g

e. Express the answer in mixed units:
14 kg 505 g − 4,288 g

f. Express the answer in mixed units:
5 kg 658 g + 57,481 g

Use a tape diagram to model each problem. Solve using a simplifying strategy or an algorithm, and write your answer as a statement.

4. One package weighs 2 kilograms 485 grams. Another package weighs 5 kilograms 959 grams. What is the total weight of the two packages?

5. Together, a pineapple and a watermelon weigh 6 kilograms 230 grams. If the pineapple weighs 1 kilogram 255 grams, how much does the watermelon weigh?

6. Javier's dog weighs 3,902 grams more than Bradley's dog. Bradley's dog weighs 24 kilograms 175 grams. How much does Javier's dog weigh?

7. The table to the right shows the weight of three Grade 4 students. How much heavier is Isabel than the lightest student?

Student	Weight
Isabel	35 kg
Irene	29 kg 38 g
Sue	29,238 g

Name ______________________________ Date ______________

1. Complete the conversion table.

Mass	
kg	**g**
1	1,000
6	
	8,000
15	
	24,000
550	

2. Convert the measurements.

a. 2 kg 700 g = __________ g

b. 5 kg 945 g = __________ g

c. 29 kg 58 g = __________ g

d. 31 kg 3 g = __________ g

e. 66,597 g = _____ kg _______ g

f. 270 kg 41 g = __________ g

3. Solve.

a. 370 g + 80 g

b. 5 kg − 730 g

c. Express the answer in the smaller unit:
27 kg 547 g + 694 g

d. Express the answer in the smaller unit:
16 kg + 2,800 g

e. Express the answer in mixed units:
4 kg 229 g − 355 g

f. Express the answer in mixed units:
70 kg 101 g − 17 kg 862 g

Use a tape diagram to model each problem. Solve using a simplifying strategy or an algorithm, and write your answer as a statement.

4. One suitcase weighs 23 kilograms 696 grams. Another suitcase weighs 25 kilograms 528 grams. What is the total weight of the two suitcases?

5. A bag of potatoes and a bag of onions combined weigh 11 kilograms 15 grams. If the bag of potatoes weighs 7 kilograms 300 grams, how much does the bag of onions weigh?

6. The table to the right shows the weight of three dogs. What is the difference in weight between the heaviest and lightest dog?

Dog	Weight
Lassie	21 kg 249 g
Riley	23 kg 128 g
Fido	21,268 g

G4-M1-SE-B1-1.3.1-1.2016

Name ______________________________ Date ______________

1. Complete the conversion table.

Liquid Capacity	
L	mL
1	1,000
5	
38	
	49,000
54	
	92,000

2. Convert the measurements.

a. 2 L 500 mL = ____________ mL

b. 70 L 850 mL = ____________ mL

c. 33 L 15 mL = ____________ mL

d. 2 L 8 mL = ____________ mL

e. 3,812 mL = _____ L _______ mL

f. 86,003 mL = _____ L _______ mL

3. Solve.

a. 1,760 mL + 40 L

b. 7 L − 3,400 mL

c. Express the answer in the smaller unit:
25 L 478 mL + 3 L 812 mL

d. Express the answer in the smaller unit:
21 L − 2 L 8 mL

e. Express the answer in mixed units:
7 L 425 mL − 547 mL

f. Express the answer in mixed units:
31 L 433 mL − 12 L 876 mL

Use a tape diagram to model each problem. Solve using a simplifying strategy or an algorithm, and write your answer as a statement.

4. To make fruit punch, John's mother combined 3,500 milliliters of tropical drink, 3 liters 95 milliliters of ginger ale, and 1 liter 600 milliliters of pineapple juice.

   a. Order the quantity of each drink from least to greatest.

   b. How much punch did John's mother make?

5. A family drank 1 liter 210 milliliters of milk at breakfast. If there were 3 liters of milk before breakfast, how much milk is left?

6. Petra's fish tank contains 9 liters 578 milliliters of water. If the capacity of the tank is 12 liters 455 milliliters of water, how many more milliliters of water does she need to fill the tank?

G4-M1-SE-B1-1.3.1-1.2016

Name ______________________________ Date ______________

1. Complete the conversion table.

Liquid Capacity	
L	**mL**
1	1,000
8	
27	
	39,000
68	
	102,000

2. Convert the measurements.

a. 5 L 850 mL = ____________ mL

b. 29 L 303 mL = ____________ mL

c. 37 L 37 mL = ____________ mL

d. 17 L 2 mL = ____________ mL

e. 13,674 mL = _____ L _____ mL

f. 275,005 mL = _____ L _____ mL

3. Solve.

a. 545 mL + 48 mL

b. 8 L − 5,740 mL

c. Express the answer in the smaller unit:
27 L 576 mL + 784 mL

d. Express the answer in the smaller unit:
27 L + 3,100 mL

e. Express the answer in mixed units:
9 L 213 mL − 638 mL

f. Express the answer in mixed units:
41 L 724 mL − 28 L 945 mL

Use a tape diagram to model each problem. Solve using a simplifying strategy or an algorithm, and write your answer as a statement.

4. Sammy's bucket holds 2,530 milliliters of water. Marie's bucket holds 2 liters 30 milliliters of water. Katie's bucket holds 2 liters 350 milliliters of water. Whose bucket holds the least amount of water?

5. At football practice, the water jug was filled with 18 liters 530 milliliters of water. At the end of practice, there were 795 milliliters left. How much water did the team drink?

6. 27,545 milliliters of gas were added to a car's empty gas tank. If the gas tank's capacity is 56 liters 202 milliliters, how much gas is needed to fill the tank?

Name ______________________________ Date ____________

1. Complete the table.

Smaller Unit	Larger Unit	How Many Times as Large as?
one	hundred	100
centimeter		100
one	thousand	1,000
gram		1,000
meter	kilometer	
milliliter		1,000
centimeter	kilometer	

2. Fill in the units in word form.

a. 429 is 4 hundreds 29 ______________.

b. 429 cm is 4 ______________ 29 cm.

c. 2,456 is 2 ______________ 456 ones.

d. 2,456 m is 2 ______________ 456 m.

e. 13,709 is 13 ______________ 709 ones.

f. 13,709 g is 13 kg 709 ______________.

3. Fill in the unknown number.

a. __________ is 456 thousands 829 ones.

b. ______________ mL is 456 L 829 mL.

4. Use words, equations, or pictures to show and explain how metric units are like and unlike place value units.

5. Compare using >, <, or =.

a. 893,503 mL ◯ 89 L 353 mL

b. 410 km 3 m ◯ 4,103 m

c. 5,339 m ◯ 533,900 cm

6. Place the following measurements on the number line:

2 km 415 m   2,379 m   2 km 305 m   245,500 cm

2,300 m   2,350 m   2,400 m   2,450 m   2,500 m

7. Place the following measurements on the number line:

2 kg 900 g   3,500 g   1 kg 500 g   2,900 g   750 g

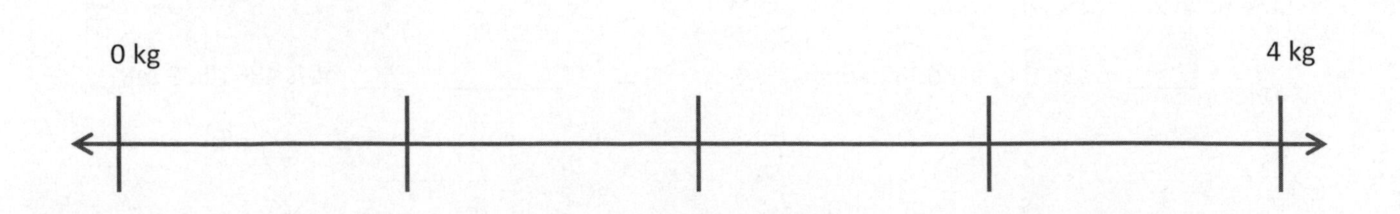

Name ______________________________ Date ______________

1. Complete the table.

Smaller Unit	Larger Unit	How Many Times as Large as?
centimeter	meter	100
	hundred	100
meter	kilometer	
gram		1,000
one		1,000
milliliter		1,000
one	hundred thousand	

2. Fill in the unknown unit in word form.

a. 135 is 1 ______________ 35 ones.

b. 135 cm is 1 ______________ 35 cm.

c. 1,215 is 1 ______________ 215 ones.

d. 1,215 m is 1 ______________ 215 m.

e. 12,350 is 12 ______________350 ones.

f. 12,350 g is 12 kg 350 ______________.

3. Write the unknown number.

a. ______________ is 125 thousands 312 ones.

b. ______________ mL is 125 L 312 mL.

4. Fill in each with >, <, or =.

   a. 890,353 mL ◯ 89 L 353 mL

   b. 2 km 13 m ◯ 2,103 m

5. Brandon's backpack weighs 3,140 grams. Brandon weighs 22 kilograms 610 grams more than his backpack. If Brandon stands on a scale wearing his backpack, what will the weight read?

6. Place the following measurements on the number line:

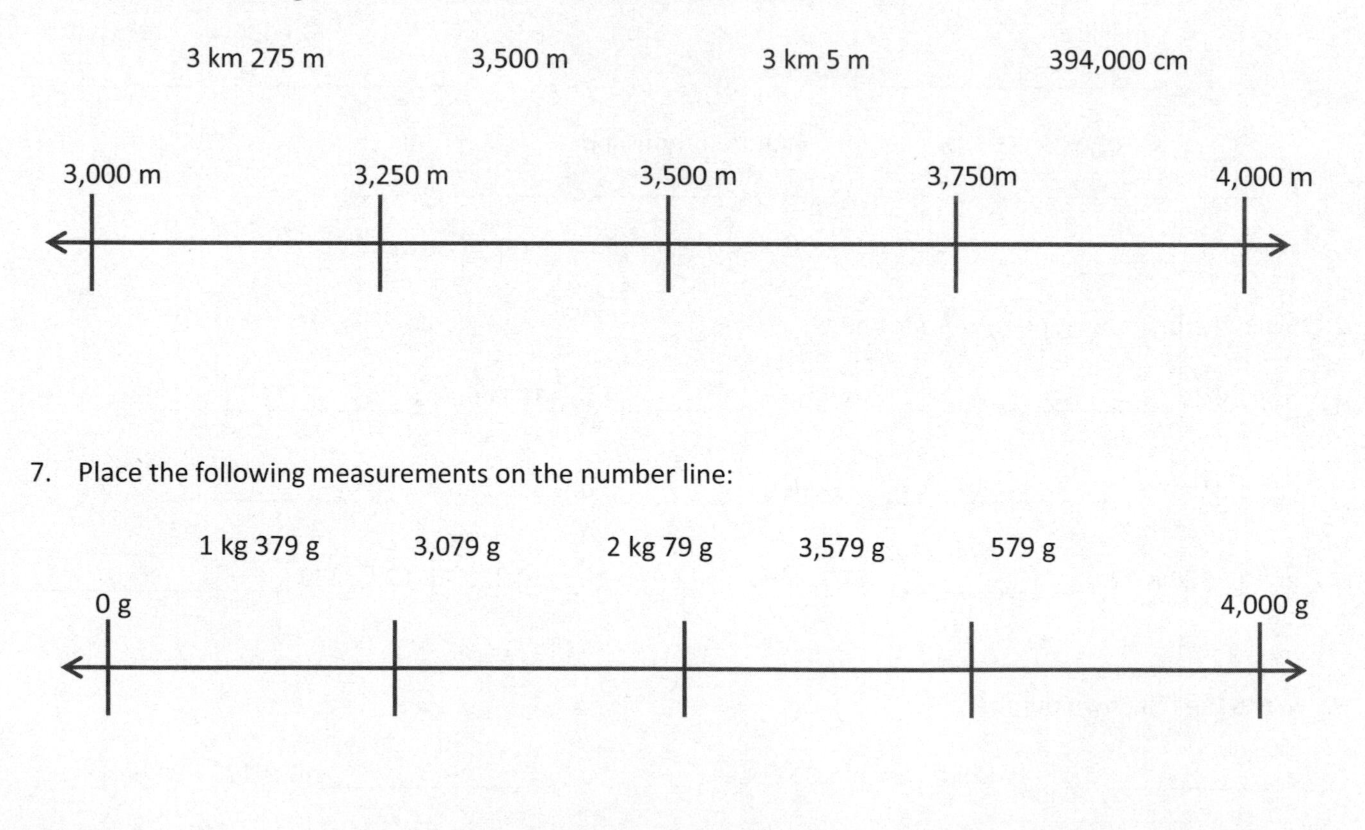

unlabeled hundred thousands place value chart

This page intentionally left blank

Name Date ____________________

Model each problem with a tape diagram. Solve and answer with a statement.

1. The potatoes Beth bought weighed 3 kilograms 420 grams. Her onions weighed 1,050 grams less than the potatoes. How much did the potatoes and onions weigh together?

2. Adele let out 18 meters 46 centimeters of string to fly her kite. She then let out 13 meters 78 centimeters more before reeling back in 590 centimeters. How long was her string after reeling it in?

3. Shyan's barrel contained 6 liters 775 milliliters of paint. She poured in 1 liter 118 milliliters more. The first day, Shyan used 2 liters 125 milliliters of the paint. At the end of the second day, there were 1,769 milliliters of paint remaining in the barrel. How much paint did Shyan use on the second day?

4. On Thursday, the pizzeria used 2 kilograms 180 grams less flour than they used on Friday. On Friday, they used 12 kilograms 240 grams. On Saturday, they used 1,888 grams more than on Friday. What was the total amount of flour used over the three days?

5. The gas tank in Zachary's car has a capacity of 60 liters. He adds 23 liters 825 milliliters of gas to the tank, which already has 2,050 milliliters of gas. How much more gas can Zachary add to the gas tank?

6. A giraffe is 5 meters 20 centimeters tall. An elephant is 1 meter 77 centimeters shorter than the giraffe. A rhinoceros is 1 meter 58 centimeters shorter than the elephant. How tall is the rhinoceros?

G4-M1-SE-B1-1.3.1-1.2016

Name ______________________________ Date ________________

Model each problem with a tape diagram. Solve and answer with a statement.

1. The capacity of Jose's vase is 2,419 milliliters of water. He poured 1 liter 299 milliliters of water into the empty vase. Then, he added 398 milliliters. How much more water will the vase hold?

2. Eric biked 1 kilometer 125 meters on Monday. On Tuesday, he biked 375 meters less than on Monday. How far did he bike both days?

3. Zachary weighs 37 kilograms 95 grams. Gabe weighs 4,650 grams less than Zachary. Harry weighs 2,905 grams less than Gabe. How much does Harry weigh?

4. A Springer Spaniel weighs 20 kilograms 490 grams. A Cocker Spaniel weighs 7,590 grams less than a Springer Spaniel. A Newfoundland weighs 52 kilograms 656 grams more than a Cocker Spaniel. What is the difference, in grams, between the weights of the Newfoundland and the Springer Spaniel?

5. Marsha has three rugs. The first rug is 2 meters 87 centimeters long. The second rug has a length 98 centimeters less than the first. The third rug is 111 centimeters longer than the second rug. What is the difference in centimeters between the length of the first rug and the third rug?

6. One barrel held 60 liters 868 milliliters of sap. A second barrel held 20,089 milliliters more sap than the first. A third barrel held 40 liters 82 milliliters less sap than the second. If the sap from the three barrels was poured into a larger container, how much sap would there be in all?